Assessment Prep for Common Core Mathematics

Grade 7

Tips and Practice for the Math Standards

AUTHORS: KARISE MACE, STEPHEN FOWLER, and CHRISTINE HENDERSON

EDITORS: MARY DIETERICH and SARAH M. ANDERSON

PROOFREADER: MARGARET BROWN

COPYRIGHT © 2015 Mark Twain Media, Inc.

ISBN 978-1-62223-530-8

Printing No. CD-404233

Mark Twain Media, Inc., Publishers
Distributed by Carson-Dellosa Publishing LLC

Table of Contents

Introduction to the Teacher

The time has come to raise the rigor in our children's mathematical education. The Common Core State Standards were developed to help guide educators and parents on how to do this by outlining what students are expected to learn at each grade level. The bar has been set high, but our students are up to the challenge.

More than 40 states have adopted the Common Core State Standards, and the school districts in those states are aligning their curriculums and state assessments to those standards. This workbook is designed to help you prepare your students for assessments based on the Common Core State Standards. It contains both multiple-choice and open-ended assessment questions that are similar to the types of questions students will encounter on their state assessments. We have also included test-taking tips and strategies that will help students perform well on these types of assessments.

Additionally, this book contains diagnostic information for the multiple-choice questions that will help you understand why your students selected particular incorrect answers. We believe that you will be able to use this information to identify the gaps in student knowledge and to inform your future instruction.

We hope that this book will be a valuable resource for you in preparing your students for assessments that are aligned with the Common Core State Standards!

— Karise Mace, Stephen Fowler, and Christine Henderson

Test-Taking Strategies for Math Tests

Test anxiety affects many students. Here are some strategies you can teach your students to help alleviate the anxiety and help them become more relaxed test-takers. We have also included a sample problem in which we highlight how these strategies might be used.

Multiple-Choice Tests

Tip #1:   Read the problem thoroughly and determine the goal.

Anxious test-takers have a tendency to read through problems quickly and then immediately scan the answer choices for what might be the correct answer. Encourage your students to be patient as they read through each problem so that they can determine what the problem is asking them to do. They may even wish to circle information that they think is important and underline the question.

Example: You have 12 yards of ribbon. It takes $\frac{2}{3}$ of a yard of ribbon to wrap a package. How many packages can you wrap?

    A. 24 packages
    B. 18 packages
    C. 16 packages
    D. 12 packages

Tip #2:   Estimate the answer.

Students often "number surf." That is, they "grab" the numbers they see in the problem and start operating on them in an attempt to get one of the answer choices. Encourage your students to use estimation to determine the reasonableness of an answer.

Estimate: I know that $\frac{2}{3}$ is more than $\frac{1}{2}$ but less than 1. So, the number of packages must be between 12 and 24.

Tip #3:   Use your estimate to quickly eliminate one or two of the choices.

Once students have calculated an estimate, they can almost always use it to eliminate one or two unreasonable choices. Encourage them to cross these out with their pencils.

Eliminate: Because the number of packages must be between 12 and 24, I can eliminate choices **A** and **D**.

Tip #4:   Solve the problem by working forward or backward.

Some problems can be solved just as efficiently by working forward or backward. If students are unsure about how to use the information in the problem to get one of the answers, encourage them to start with one of the answers and work backward to see if they get the information in the problem.

Working forward:

12 yards $\div \frac{2}{3}$ yard/package =
    $\frac{12}{1} \times \frac{3}{2}$ = 18 packages

The correct answer is choice **B**.

Test-Taking Strategies for Math Tests

Open-Ended Response Tests

The tips for solving open-ended response problems are similar to those for solving multiple-choice problems. However, because open-ended response questions are also used to assess the problem-solving process, students must learn how to communicate their process. These tips will help them learn to do that.

Tip #1:   Read the problem thoroughly and determine the goal.

Open-ended response problems are often multi-step. It is important to encourage your students to read these problems patiently and thoroughly so that they do not forget to complete the problem. It may be helpful for them to circle important information and underline the question.

Example: Maggie has 110 feet of fencing and would like to use it to enclose a rectangular area that is 32 feet by 25 feet. Does she have enough fencing to do this? Explain your reasoning.

Tip #2:   Make a list of what you know and what you need to figure out.

Making lists can help students keep their information organized. Encourage them to make two lists—one of the things they know and another of the things they need to figure out.

Things I know:	**Things I need to figure out:**
1. Maggie has 110 feet of fencing.	1. What is the perimeter of the area to be enclosed?
2. The area to be enclosed is a rectangle.	
3. The length of the rectangle is 32 feet, and the width is 25 feet.	2. Whether or not Maggie has enough fencing to enclose the area

Tip #3:   Devise a plan for solving the problem.

While students do not always need to write out their problem-solving plan, it is important for them to form one. Many open-ended response problems ask students to explain their problem-solving process. Encourage students to write down their plan as part of this explanation.

Plan: I am going to calculate the perimeter of the rectangular area and compare it to the amount of fencing Maggie has.

Tip #4:   Carry out your plan.

As students begin to carry out their plan, encourage them to show their work!

Tip #5:   Check your work.

Students like to skip this step, but it is one of the most important ones in the problem-solving process. Encourage your students to take time to check their work and to make sure that they actually solved the problem they were asked to solve.

Carry out the plan:

$P = 2l + 2w$

$\quad = 2(32) + 2(25)$

$\quad = 114$

The perimeter of the rectangular area is 114 feet. Maggie does *not* have enough fencing to enclose it because she only has 110 feet of fencing.

Geometry

Problem Correlation to CCSS Grade 7 Geometry Standards

MC Problem #	7.G.A.1	7.G.A.2	7.G.A.3	7.G.B.4	7.G.B.5	7.G.B.6
1			•			
2				•		
3	•					
4		•				
5						•
6					•	
7	•					
8		•				
9			•			
10				•		
11					•	
12						•
13					•	
14				•		
15			•			
16		•				
17	•					
18				•		
19						•
20						•
Open-Ended Problem #	7.G.A.1	7.G.A.2	7.G.A.3	7.G.B.4	7.G.B.5	7.G.B.6
1				•		
2	•					
3						•
4					•	

Name: _____ Date: _____

Geometry: Multiple-Choice Assessment Prep

Directions: Circle the choice that best answers the question.

1. Which answer best describes the two-dimensional shape obtained by horizontally slicing the right hexagonal prism as shown?

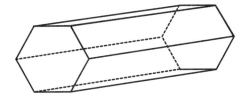

 A. Trapezoid

 B. Hexagon

 C. Rectangle

 D. Parallelogram

3. The rectangle shown below is scaled by a factor of $\frac{2}{3}$. What is the area of the new rectangle?

6 | [   ] |
24

 A. 64

 B. 40

 C. 96

 D. 324

2. The midfield circle on a soccer field has a diameter of 60 ft. Esteban runs once around the midfield circle. To the nearest foot, how far does Esteban run?

 A. 188 ft

 B. 377 ft

 C. 2,826 ft

 D. 60 ft

4. How many unique triangles can be constructed using angles measuring 40°, 60°, and 80°?

 A. None

 B. One

 C. Two

 D. Infinitely many

Name: _____ Date: _____

Geometry: Multiple-Choice Assessment Prep

Directions: Circle the choice that best answers the question.

---

5. A solid number cube made of plastic has an edge measuring $\frac{5}{8}$ in. How much plastic is needed to make the number cube?

A. $\frac{75}{32}$ in.³

B. $\frac{75}{32}$ in.²

C. $\frac{25}{64}$ in.²

D. $\frac{125}{512}$ in.³

---

7. What factor is used to scale the triangle shown below?

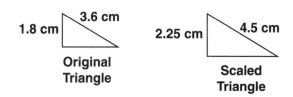

Original Triangle   Scaled Triangle

A. 1.25

B. 0.8

C. 2

D. 0.5

---

6. Determine the measure of $\angle APD$:

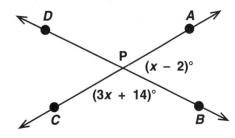

A. 40°

B. 140°

C. 42°

D. 158°

---

8. One side of a triangle measures 7.5 units. Another side measures 11.6 units. The third side has a length that is a whole number of units. What is the smallest possible length of the third side?

A. 4 units

B. 19 units

C. 5 units

D. 4.1 units

Name: _____   Date: _____

Geometry: Multiple-Choice Assessment Prep

Directions: Circle the choice that best answers the question.

---

9. The right square pyramid below is sliced vertically. Which of the following figures cannot be the two-dimensional shape obtained?

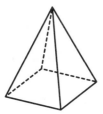

   A. Trapezoid

   B. Square

   C. Triangle

   D. Any of these shapes is possible

---

11. Quadrilateral *MATH* is a rectangle. Determine the measure of $\angle HAT$.

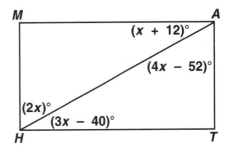

   A. 26°

   B. $\left(33\frac{1}{3}\right)°$

   C. 38°

   D. 52°

---

10. Jin opens a shade umbrella with a diameter of $6\frac{1}{2}$ feet to sit beneath while at the beach. What is the best approximation of the area of sand shaded by Jin's umbrella when the sun is directly overhead?

   A. 33 square feet

   B. 133 square feet

   C. 20 square feet

   D. 11 square feet

---

12. A feeding trough is in the shape of a right triangular prism. The triangular end of the trough measures 3 ft across and is $1\frac{1}{4}$ ft deep. The feeding trough is 40 feet long. How much cattle feed can the trough hold?

   A. 150 ft³

   B. 75 ft³

   C. 300 ft³

   D. 120 ft²

Name: _____ Date: _____

Geometry: Multiple-Choice Assessment Prep

Directions: Circle the choice that best answers the question.

13. Determine the measure of ∠EVF:

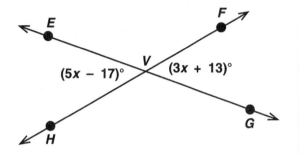

A. 58°

B. 122°

C. 15°

D. 98°

---

15. The trunk of a fallen tree lies on the ground as shown below. Clayton cuts the log to obtain a circular piece to use as the seat of a stool. In what direction does Clayton cut the fallen tree?

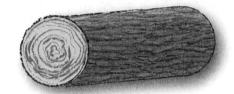

A. Diagonally

B. Horizontally

C. Vertically

D. Clayton cannot make the necessary cut.

---

14. Maria has a circular flower garden in her backyard. The stone path around the garden measures 1,413 cm. To the nearest thousand, what is the area of Maria's flower garden?

A. 499,000 cm²

B. 256,000 cm²

C. 636,000 cm²

D. 159,000 cm²

---

16. Which of the following sets of conditions can be used to construct a triangle?

A. Sides measuring 28, 46, and 74 units

B. Three obtuse angles

C. Three sides measuring 180 cm each

D. Two right angles

Name: _____   Date: _____

Geometry: Multiple-Choice Assessment Prep

Directions: Circle the choice that best answers the question.

---

17. A small airport expands its rectangular runway by a scale factor of $\frac{3}{2}$. The new runway is $\frac{3}{4}$ mi long and covers $\frac{1}{24}$ mi². What is the width of the original runway?

A. $\frac{1}{27}$ mi

B. $\frac{1}{12}$ mi

C. $\frac{1}{48}$ mi

D. $\frac{3}{64}$ mi

---

18. Nelson's bicycle has 26-inch diameter wheels. Nelson pedals his bicycle enough for the wheels to complete exactly 250 revolutions. To the nearest hundred, how far does Nelson travel?

A. 132,700 inches

B. 20,400 inches

C. 6,500 inches

D. 40,800 inches

---

19. Ye builds a plywood storage box in the shape of a right trapezoidal prism as shown below. How much plywood does Ye use?

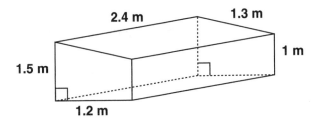

A. 3.6 m³

B. 4.32 m³

C. 15.75 m²

D. 15 m²

---

20. An aerial view of a public swimming pool is shown below. Which is the best approximation of its area?

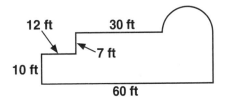

A. 1,445 ft²

B. 1,147 ft²

C. 1,063 ft²

D. 757 ft²

Name: _____ Date: _____

Geometry: Open-Ended Response Assessment Prep

Directions: Answer the question completely. Show your work and explain your reasoning.

Problem 1: A pizza shop sells its medium cheese pizza for $8.95 and its large cheese pizza for $12.95. The diameters of the pizzas are 12 inches and 16 inches, respectively. Which is the better deal?

Show your work.	**Explain your reasoning.**

Name: _____ Date: _____

Geometry: Open-Ended Response Assessment Prep

Directions: Answer the question completely. Show your work and explain your reasoning.

Problem 2: The scale drawing below shows the route April drives to her cousin's house. In the figure, one inch represents $2\frac{1}{2}$ miles. How far does April drive to visit her cousin?

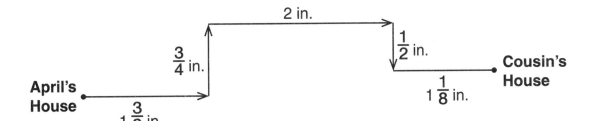

Show your work.

Explain your reasoning.

Name: _____ Date: _____

Geometry: Open-Ended Response Assessment Prep

Directions: Answer the question completely. Show your work and explain your reasoning.

Problem 3: The competition swimming pool shown below is resurfaced with quartz aggregate that costs $7.50 per square foot. What is the total cost of resurfacing the walls and floor of the pool?

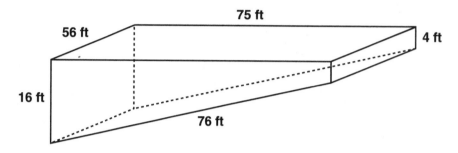

Show your work.

Explain your reasoning.

Name: _____ Date: _____

Geometry: Open-Ended Response Assessment Prep

Directions: Answer the question completely. Show your work and explain your reasoning.

Problem 4: Determine the measure of ∠*BPC* in the figure below.

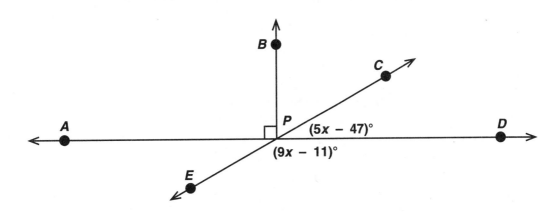

Show your work.	Explain your reasoning.

Geometry: Answers and Diagnostics

Multiple Choice Questions

Problem #	Correct Answer	Diagnostics
1.	C	A. Student responded with front view of remaining shape after horizontal slicing. B. Student responded with answer after vertical slicing. D. Student did not provide best possible response.
2.	A	B. Student interpreted given diameter as radius. C. Student calculated area instead of circumference. D. Student did not understand concept of diameter or circumference.
3.	A	B. Student calculated perimeter of new rectangle. C. Student multiplied area of given rectangle by $\frac{2}{3}$. D. Student multiplied by $\frac{3}{2}$ instead of $\frac{2}{3}$.
4.	D	A. Student added angles incorrectly to obtain sum not equal to 180°. B. Student incorrectly concluded scaling side lengths also scales angle measures. C. Student did not consider all possible side lengths.
5.	D	A. Student calculated surface area and used incorrect units for surface area. B. Student calculated surface area. C. Student calculated area of one face of number cube.
6.	B	A. Student incorrectly equated $\angle APD$ to $\angle APB$ instead of $\angle CPB$. C. Student only solved for x. D. Student incorrectly added 12 to solve correct equation ($4x + 12 = 180$).
7.	A	B. Student divided original side by scaled side. C. Student divided longer side by shorter side of the same triangle. D. Student divided shorter side by longer side of the same triangle.
8.	C	A. Student rounded down to 4 units, but side must be greater than 4.1 units. B. Student added given side lengths instead of subtracting. D. Student gave exact value rather than whole number as directed.
9.	B	A. Student did not consider slicing down a face, cutting two edges. C. Student did not consider slicing through only one edge or through the apex. D. Student incorrectly considered slicing horizontally to obtain a square.
10.	A	B. Student used diameter instead of radius when calculating area. C. Student calculated circumference and used incorrect units. D. Student did not multiply by π when calculating area.
11.	D	A. Student only solved for x. B. Student incorrectly equated $\angle HAT$ to $\angle HAM$ instead of summing to 90°. C. Student gave measure of $\angle HAM$ instead of $\angle HAT$.
12.	B	A. Student did not multiply by $\frac{1}{2}$ when calculating area of triangular base. C. Student multiplied by 2 rather than $\frac{1}{2}$ when calculating area of triangular base. D. Student calculated area of top of trough instead of volume.
13.	B	A. Student solved for either $\angle EVH$ or $\angle FVG$. C. Student only solved for x. D. Student summed $\angle EVH$ and $\angle FVG$ to 180° instead of equating them, then used $\angle FVG$ to determine measure of $\angle EVF$.
14.	D	A. Student calculated area as $(\pi r)^2$ instead of πr^2. B. Student guessed. C. Student used diameter instead of radius when calculating area.

Geometry: Answers and Diagnostics

15.	C	A. Student did not properly take perspective of drawing into account. B. Student confused the concepts of horizontal and vertical. D. Student did not understand the question.
16.	C	A. Student incorrectly concluded sum of two smaller sides equal to longest side forms a valid triangle (sum of two smaller sides must be less than longest side). B. Student misinterpreted obtuse to mean acute. D. Student did not consider third angle of triangle to obtain sum of 180°.
17.	A	B. Student multiplied width of new runway by $\frac{3}{2}$ instead of dividing by $\frac{3}{2}$. C. Student multiplied area by length to obtain width of new runway before dividing by correct scale factor of $\frac{3}{2}$. D. Student multiplied area by length to obtain width of new runway instead of dividing, then multiplied width of new runway by $\frac{3}{2}$ instead of dividing by $\frac{3}{2}$.
18.	B	A. Student calculated area instead of circumference. C. Student multiplied diameter, not circumference, by 250. D. Student interpreted given diameter as radius.
19.	D	A. Student calculated volume instead of surface area. B. Student calculated volume instead of surface area and calculated area of trapezoidal base as if it were a rectangle. C. Student used top and bottom edges as bases of trapezoid, not parallel edges.
20.	C	A. Student used diameter instead of radius for area of semicircle. B. Student did not subtract rectangular cutout in top left corner of figure. D. Student used 30 for length of larger rectangular region.

Open-Ended Response Questions

Problem #1
Medium: Area = 3.14(6)^2 = 113.04 sq in.; cost per sq in. = 8.95 ÷ 113.04 ≈ $0.079 = 7.9¢
Large: Area = 3.14(8)^2 = 200.96 sq in.; cost per sq in. = 12.95 ÷ 200.96 ≈ $0.064 = 6.4¢
The large pizza is the better deal by approximately 1.5¢ per square inch.

Problem #2

Total length of path in scale drawing: $1\frac{3}{8} + \frac{3}{4} + 2 + \frac{1}{2} + 1\frac{1}{8} = \frac{11}{8} + \frac{6}{8} + \frac{16}{8} + \frac{4}{8} + \frac{9}{8} = \frac{46}{8} = 5\frac{3}{4}$ in.

Total distance traveled: $\left(5\frac{3}{4}\right)\left(2\frac{1}{2}\right) = \left(\frac{23}{4}\right)\left(\frac{5}{2}\right) = \frac{115}{8} = 14\frac{3}{8}$ mi; April drives $14\frac{3}{8}$ miles to visit her cousin.

Problem #3

Area of each trapezoidal side wall: $\frac{1}{2}(16 + 4)(75) = \frac{1}{2}(20)(75) = 10(75) = 750$
Deep wall: 16(56) = 896; shallow wall: 4(56) = 224; floor: 56(76) = 4,256
Total area to be resurfaced: 2(750) + 896 + 224 + 4,256 = 1,500 + 896 + 224 + 4,256 = 6,876 sq ft
Total cost: 6,876(7.50) = $51,570
The total cost to resurface the competition swimming pool is $51,570.

Problem #4
$m\angle CPD + m\angle EPD = 180 \rightarrow (5x - 47) + (9x - 11) = 180 \rightarrow 14x - 58 = 180 \rightarrow 14x = 238 \rightarrow$
$x = 17$
$m\angle CPD = 5x - 47 = 5(17) - 47 = 85 - 47 = 38°$; $m\angle BPC = 90 - m\angle CPD = 90 - 38 = 52°$
The measure of $\angle BPC$ is 52°.

Ratios and Proportional Relationships

Problem Correlation to CCSS Grade 7 Ratios and Proportional Relationships Standards

MC Problem #	7.RP.A.1	7.RP.A.2	7.RP.A.2a	7.RP.A.2b	7.RP.A.2c	7.RP.A.2d	7.RP.A.3
1	•						
2							•
3					•		
4			•				
5				•			
6							•
7						•	
8					•		
9	•						
10			•				
11				•			
12			•				
13						•	
14	•						
15							•
16					•		
17				•			
18							•
19						•	
20	•						
Open-Ended Problem #	7.RP.A.1	7.RP.A.2	7.RP.A.2a	7.RP.A.2b	7.RP.A.2c	7.RP.A.2d	7.RP.A.3
1							•
2					•		
3	•						
4			•				

Name: _____   Date: _____

Ratios and Proportional Relationships: Multiple-Choice Assessment Prep

Directions: Circle the choice that best answers the question.

1. A cookie recipe calls for $\frac{2}{3}$ cup brown sugar and $\frac{1}{4}$ teaspoon salt. How many cups of brown sugar per teaspoon of salt are needed?

    A. $\frac{3}{8}$ cup per teaspoon

    B. 8 cups per 3 teaspoons

    C. $\frac{1}{6}$ cup per teaspoon

    D. $\frac{8}{3}$ cups per teaspoon

2. A computer regularly priced at $843.75 is marked down 40%. What is the sale price of the computer?

    A. $803.75

    B. $506.25

    C. $337.50

    D. $1,181.25

3. The total height T of a stack of blocks is proportional to the number n of blocks with a constant height h. Which is the correct equation to represent the situation?

    A. $T = hn$

    B. $h = Tn$

    C. $T = \dfrac{h}{n}$

    D. $h = \dfrac{n}{T}$

4. Are the quantities x and y proportional?

x	1	2	3
y	1	4	9

    A. Yes

    B. No

    C. Answer depends on the meaning of x and y.

    D. Not enough information

Ratios and Proportional Relationships: Multiple-Choice Assessment Prep

Directions: Circle the choice that best answers the question.

5. Cornell runs a 5 km race. The graph shows his progress. What is Cornell's rate in kilometers per minute?

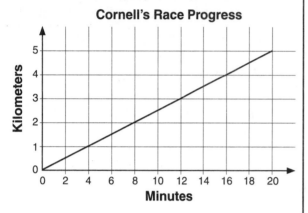

Cornell's Race Progress

A. 4 kilometers per minute

B. 0.5 kilometer per minute

C. 0.25 kilometer per minute

D. 2 kilometers per minute

6. Pavel estimates he can average 11.2 points per game this basketball season. His actual average is 12.8 points per game. To the nearest tenth, what is the percent error in Pavel's estimate?

A. 12.5%

B. 1.6%

C. 87.5%

D. 14.3%

7. Explain the meaning of the point labeled in the graph.

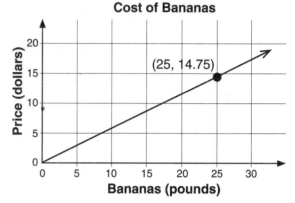

Cost of Bananas

A. The maximum amount of bananas available for purchase is 25 pounds, and they will cost $14.75.

B. It costs $25.00 for 14.75 pounds of bananas.

C. The cost of 25 bananas is $14.75.

D. Twenty-five pounds of bananas cost $14.75.

8. The weight of water is proportional to its volume. Determine the equation for the weight W of g gallons of water given that water weighs $\frac{25}{3}$ pounds per gallon.

A. $g = \frac{25}{3}W$

B. $W = \frac{25}{3g}$

C. $W = \frac{25}{3}g$

D. Cannot be determined

Name: _____   Date: _____

Ratios and Proportional Relationships: Multiple-Choice Assessment Prep

Directions: Circle the choice that best answers the question.

---

9. The scale on a city map is $\frac{3}{4}$ inch represents $\frac{1}{8}$ mile. How many inches per mile are represented by the scale on the map?

   A. $\frac{3}{32}$ inch per mile

   B. 6 inches per mile

   C. $\frac{1}{6}$ inch per mile

   D. $\frac{32}{3}$ inches per mile

---

11. Honey is produced proportionally to the number of bees in a colony. A large colony of 96,000 honeybees produces 2,000 cups of honey. How many cups of honey per honeybee are produced?

   A. $\frac{1}{48}$ cup per honeybee

   B. 192,000,000 cups per honeybee

   C. $\frac{5}{24}$ cup per honeybee

   D. 48 cups per honeybee

---

10. Which of the following sets of values represents a proportional relationship?

Option I		
A	0	2	6
B	0	5	15

Option II		
P	1	5	10
Q	3.2	16	32

   A. I only

   B. II only

   C. Both I and II

   D. Neither I nor II

---

12. Hideki earns 600 bonus points for unlocking 4 secret doors in a video game. Francisco earns 1,050 bonus points for unlocking 7 secret doors in the same game. Briana earns 1,800 bonus points for unlocking 12 secret doors. Is the number of bonus points proportional to the number of secret doors unlocked?

   A. Yes

   B. No

   C. Answer depends on number of secret doors unlocked

   D. Cannot be determined

Name: _____ Date: _____

Ratios and Proportional Relationships: Multiple-Choice Assessment Prep

Directions: Circle the choice that best answers the question.

13. The graph below shows Xin's pay earned at work. Based on the information provided, how much does Xin earn per hour?

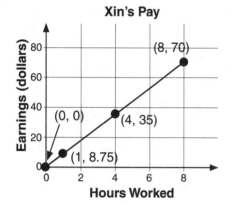

A. $70.00

B. $35.00

C. $8.00

D. $8.75

14. A cheetah runs $1\frac{2}{3}$ meters in 0.1 second. Determine the cheetah's speed in meters per second.

A. $\frac{1}{6}$ meter per second

B. $16\frac{2}{3}$ meters per second

C. $\frac{3}{50}$ meter per second

D. $6\frac{2}{3}$ meters per second

15. For large parties, restaurants often include a tip on the final bill that is proportional to the total amount spent. One restaurant includes a tip of $24.21 for a large family that spends $134.50 on dinner. What is the amount of the tip the restaurant includes for a large group that spends $246.00 on dinner?

A. $42.21

B. $18.00

C. $290.28

D. $44.28

16. The total distance d traveled on a bike is proportional to the number of revolutions R of its wheels with circumference C. Which is the correct equation to represent this situation?

A. $d = \frac{R}{C}$

B. $R = \frac{C}{d}$

C. $d = CR$

D. $C = dR$

Name: _____ Date: _____

Ratios and Proportional Relationships: Multiple-Choice Assessment Prep

Directions: Circle the choice that best answers the question.

---

17. Given that y is proportional to x, use the table below to determine the constant of proportionality k in the equation $y = kx$.

x	0	2.8	5.6	11.2
y	0	8.75	17.5	35

A. 2

B. 0.32

C. 3.125

D. There is no constant of proportionality.

---

19. The graph below shows information about raffle tickets sold at a charity event. Explain the meaning of the point labeled on the graph.

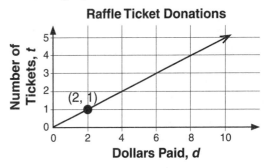

Raffle Ticket Donations

A. Each dollar buys 2 raffle tickets.

B. Each raffle ticket costs $2.00.

C. Each person buys 2 raffle tickets.

D. There is not enough information provided.

---

18. Since beginning a regular exercise routine, Anna eats 2,700 calories each day, which is a 20% increase over her daily caloric intake prior to beginning the exercise routine. How many calories did Anna consume each day before participating in her regular exercise routine?

A. 2,250 calories

B. 3,240 calories

C. 2,160 calories

D. 2,680 calories

---

20. Chantelle's heart beats 63 times for every 14 breaths she takes while sleeping. How many times does Chantelle's heart beat per breath?

A. 50 beats per breath

B. $\frac{2}{9}$ beat per breath

C. 882 beats per breath

D. 4.5 beats per breath

Name: _____ Date: _____

Ratios and Proportional Relationships: Open-Ended Response Assessment Prep

Directions: Answer the question completely. Show your work and explain your reasoning.

Problem 1: Triangle *XYZ* is proportional to triangle *ABC*. The perimeter of triangle *XYZ* is 18 ft. Determine the length of $\overline{XZ}$.

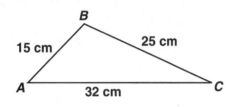

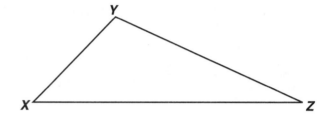

Show your work.

Explain your reasoning.

Name: _____   Date: _____

Ratios and Proportional Relationships: Open-Ended Response Assessment Prep

Directions: Answer the question completely. Show your work and explain your reasoning.

Problem 2: The length L of a shadow cast by the sun at a particular time of day is proportional to the height h of the object producing the shadow. Use the table below to write an equation to represent this situation. Then determine the length of a shadow cast by a building 123 feet tall using your equation.

h (ft)	L (ft)
0	0
6	4.32
25	18
46	33.12

Show your work.

Explain your reasoning.

Name: _____ Date: _____

Ratios and Proportional Relationships: Open-Ended Response Assessment Prep

Directions: Answer the question completely. Show your work and explain your reasoning.

Problem 3: A broken pipe leaks $5\frac{1}{2}$ gallons of water every $\frac{1}{4}$ hour. To the nearest hundredth, determine the rate at which the broken pipe leaks water in gallons per minute.

Show your work.

Explain your reasoning.

Name: _____   Date: _____

Ratios and Proportional Relationships:
Open-Ended Response Assessment Prep

Directions: Answer the question completely. Show your work and explain your reasoning.

Problem 4: A 3-pound bag of gourmet coffee beans costs $53.97. A 10-pound bag of the same gourmet coffee beans costs $179.90. Is the price of these gourmet coffee beans proportional to their weight?

Show your work.

Explain your reasoning.

Ratios and Proportional Relationships: Answers and Diagnostics

Multiple Choice Questions

Problem #	Correct Answer	Diagnostics
1.	D	A. Student reported teaspoons of salt per cup of brown sugar and labeled with incorrect units. B. Student divided correctly but reported as ratio rather than unit rate as directed. C. Student multiplied instead of dividing.
2.	B	A. Student subtracted $40 instead of subtracting 40% of original price. C. Student calculated amount of markdown rather than price after markdown. D. Student added markdown instead of subtracting it.
3.	A	B. Student confused total height and constant height of each block. C. Student divided height of block by number of blocks instead of multiplying. D. Student guessed.
4.	B	A. Student calculated ratios of y to x incorrectly. C. Student did not understand general nature of proportional relationships. D. Student did not understand how to verify proportional relationships.
5.	C	A. Student divided minutes by kilometers rather than kilometers by minutes. B. Student counted tick marks on graph but did not consider units. D. Student divided minutes by kilometers and did not consider units.
6.	A	B. Student reported amount of error incorrectly labeled as percentage. C. Student compared estimate rather than amount of error to actual value. D. Student compared amount of error to estimate rather than to actual value.
7.	D	A. Student misinterpreted labeled point as endpoint of graph. B. Student confused horizontal and vertical axes. C. Student did not consider units for bananas sold.
8.	C	A. Student confused weight and volume. B. Student divided constant of proportionality by g instead of multiplying by g. D. Student did not understand the question.
9.	B	A. Student multiplied instead of dividing. C. Student divided miles by inches rather than inches by miles. D. Student guessed.
10.	C	A. Student miscalculated constant of proportionality for II. B. Student miscalculated constant of proportionality for I. D. Student miscalculated constants of proportionality.
11.	A	B. Student multiplied quantities instead of dividing. C. Student divided by 9,600 rather than 96,000. D. Student calculated unit rate of honeybees to honey and labeled incorrectly.
12.	A	B. Student calculated unit rates of points per secret door incorrectly. C. Student did not understand general nature of proportional relationships. D. Student did not understand how to verify proportional relationships.
13.	D	A. Student reported earnings for entire 8-hour workday. B. Student guessed. C. Student reported number of hours worked with incorrect units.
14.	B	A. Student multiplied by 0.1 instead of dividing. C. Student calculated unit rate in seconds per meter and labeled incorrectly. D. Student divided $\frac{2}{3}$ by 0.1 instead of dividing $1\frac{2}{3}$ by 0.1.

Ratios and Proportional Relationships: Answers and Diagnostics

15.	D	A. Student calculated correct tip rate of 18% but then added 18 to original tip. B. Student incorrectly reported tip percentage as tip amount in dollars. C. Student reported total amount paid rather than only tip amount as requested.
16.	C	A. Student divided by constant of proportionality C instead of multiplying by C. B. Student guessed. D. Student confused circumference and total distance traveled.
17.	C	A. Student divided consecutive values of x or y instead of dividing y by x. B. Student divided x by y instead of dividing y by x. D. Student divided by 0 using first ordered pair or did not understand question.
18.	A	B. Student increased current number of calories by 20%. C. Student decreased current number of calories by 20%. D. Student subtracted 20 calories from current number of calories.
19.	B	A. Student confused meaning of horizontal and vertical axes. C. Student incorrectly interpreted meaning of horizontal axis. D. Student did not understand the question.
20.	D	A. Student subtracted 13 from 14 to obtain 1 breath, then likewise subtracted 13 from 63, instead of dividing 63 by 14. B. Student divided breaths by heartbeats rather than heartbeats by breaths. C. Student multiplied instead of dividing.

Open-Ended Response Questions

Problem #1

Perimeter of triangle ABC: $15 + 25 + 32 = 72$ cm

$$\frac{\text{Perimeter of triangle } ABC}{\text{Perimeter of triangle } XYZ} = \frac{\text{Length of } AC}{\text{Length of } XZ} \rightarrow \frac{72 \text{ cm}}{18 \text{ ft}} = \frac{32 \text{ cm}}{x \text{ ft}} \rightarrow 72x = 576 \rightarrow x = 8 \text{ ft.}$$

The length of $\overline{XZ}$ is 8 ft.

Problem #2

Constant of proportionality (i.e., unit rate of shadow length per foot of object height): $\frac{18}{25} = 0.72$
Equation representing situation: $L = 0.72h$; 123-foot tall building means $h = 123$; length of shadow cast by building: $L = 0.72(123) = 88.56$ feet; The desired equation is $L = 0.72h$, and the shadow cast by the 123-foot tall building is 88.56 feet long.

Problem #3

Unit rate in gallons per hour: $\frac{5\frac{1}{2}}{\frac{1}{4}} = 5\frac{1}{2} \div \frac{1}{4} = \frac{11}{2} \times \frac{4}{1} = \frac{44}{2} = 22$ gallons per hour

Unit rate in gallons per minute: $22 \div 60 = 0.3\overline{6} \approx 0.37$
The broken pipe leaks approximately 0.37 gallon of water per minute.

Problem #4

Unit price of 3-pound bag: $53.97 \div 3 = \$17.99$ per pound; Unit price of 10-pound bag: $\$179.90 \div 10 = \$17.99$ per pound; Unit prices are the same, so the price is proportional to weight.
The price of the gourmet coffee beans is proportional to their weight.

The Number System

Problem Correlation to CCSS Grade 7 The Number System Standards

MC Problem #	7.NS.A.1	7.NS.A.1a	7.NS.A.1b	7.NS.A.1c	7.NS.A.1d	7.NS.A.2	7.NS.A.2a	7.NS.A.2b	7.NS.A.2c	7.NS.A.2d	7.NS.A.3
1					•						
2				•							
3			•								
4		•									
5										•	
6									•		
7								•			
8							•				
9											•
10				•							
11									•		
12		•									
13							•				
14											•
15								•			
16			•								
17					•						
18										•	
19											•
20											•
Open-Ended Problem #	7.NS.A.1	7.NS.A.1a	7.NS.A.1b	7.NS.A.1c	7.NS.A.1d	7.NS.A.2	7.NS.A.2a	7.NS.A.2b	7.NS.A.2c	7.NS.A.2d	7.NS.A.3
1	•										
2						•					
3											•
4											•

Name: _____ Date: _____

The Number System: Multiple-Choice Assessment Prep

Directions: Circle the choice that best answers the question.

1. Determine the simplified fractional form of the following expression:

    $$\frac{17}{40} - 3.14 + 2$$

    A. $-\frac{143}{200}$

    B. -0.715

    C. $-4\frac{143}{200}$

    D. $-\frac{715}{1000}$

2. At the start of a play during a football game, Marshawn is $7\frac{1}{4}$ yards behind the line of scrimmage as the running back, and Russell is $4\frac{3}{8}$ yards behind the line of scrimmage as an offensive lineman. What is the difference in their distances from the line of scrimmage?

    A. $11\frac{5}{8}$ yd

    B. $3\frac{1}{8}$ yd

    C. $1\frac{1}{8}$ yd

    D. $2\frac{7}{8}$ yd

3. Linda walks $5\frac{1}{2}$ miles due north from her house. She then walks $2\frac{1}{3}$ miles due south. Where is Linda relative to her starting point?

    A. $3\frac{1}{6}$ mi south

    B. $3\frac{1}{6}$ mi north

    C. $7\frac{5}{6}$ mi north

    D. $7\frac{5}{6}$ mi south

4. Noriko has $37.62. She lends $19.35 to Chin. How can Noriko get back to her starting amount?

    A. Buy a board game that costs $18.27

    B. Purchase soccer cleats that cost $56.97

    C. Earn $19.35 doing chores

    D. Donate $19.35 to charity

Name: _____ Date: _____

The Number System: Multiple-Choice Assessment Prep

Directions: Circle the choice that best answers the question.

---

5. Convert $\frac{2}{41}$ to a decimal using long division.

   A. $0.\overline{4878}$

   B. $0.0\overline{4878}$

   C. $0.04\overline{87}$

   D. 0.04878

---

7. Emily bakes cherry pies to sell at her bakery. Each pie requires $\frac{2}{3}$ cup brown sugar. How many pies can she bake if she has $4\frac{1}{2}$ cups of brown sugar?

   A. 7

   B. 3

   C. 6

   D. $3\frac{5}{6}$

---

6. Determine the simplified decimal form of the following expression:

$$36 \div \frac{4}{9} \times 2.7$$

   A. 218.7

   B. $\frac{2187}{10}$

   C. 30

   D. 43.2

---

8. Eduardo's family sets a goal to decrease the amount of garbage they produce this year by $\frac{1}{5}$ ton. If they decrease their garbage production by $1\frac{3}{4}$ times their goal amount, how much less garbage will Eduardo's family produce this year?

   A. 1.95 T

   B. 8.75 T

   C. 1.55 T

   D. 0.35 T

Name: _____ Date: _____

The Number System: Multiple-Choice Assessment Prep

Directions: Circle the choice that best answers the question.

9. A woman runs $2\frac{1}{2}$ miles each day, five days per week, for exercise. To avoid getting stuck in a rut, she decreases the amount she runs each day by 15%, but she runs one additional day each week. How does the distance she runs each week under her new routine compare to her previous routine?

   A. She now runs $\frac{1}{4}$ mi. more each week.

   B. She now runs $4\frac{3}{4}$ mi. more each week.

   C. She now runs $1\frac{3}{5}$ mi. more each week.

   D. She now runs $\frac{1}{4}$ mi. less each week.

11. Completely simplify the following expression:

$$\frac{\frac{9}{8} \times \frac{4}{3}}{\frac{1}{2} \div \frac{3}{4}}$$

   A. 1

   B. 4

   C. $2\frac{1}{4}$

   D. $\frac{1}{16}$

10. The left fielder on a softball team is standing 87 feet to the left of the center fielder, and the right fielder is standing 134 feet to the right of the center fielder. How far apart are the left fielder and right fielder?

   A. 221 ft

   B. 47 ft

   C. 110.5 ft

   D. Cannot be determined

12. Each magnesium ion has an electrical charge of positive two. Each phosphate ion has an electrical charge of negative three. Which of these combinations has a net charge of zero?

   A. Four magnesium and six phosphate ions

   B. Six magnesium and four phosphate ions

   C. Two magnesium and three phosphate ions

   D. Five magnesium and five phosphate ions

Name: _____ Date: _____

The Number System: Multiple-Choice Assessment Prep

Directions: Circle the choice that best answers the question.

13. Taisha works a part-time job while taking evening classes at her local community college. She sets up a payment plan so that she pays $320.67 per month to attend school. If Taisha earns $292.36 per month, which value best represents her financial status after one year?

A. $3,508.32

B. $339.72

C. –$339.72

D. –$3,848.04

14. Forty percent of the difference between 18.9 and –23.7 is scaled by a factor of $\frac{5}{8}$. That result is then evenly divided into thirds. What is the value of each of those thirds?

A. 9.088

B. –0.4

C. 0.355

D. 3.55

15. A 12-inch ruler is divided into 48 equal parts. What is the size of each part?

A. 4 in.

B. 1 in.

C. $\frac{1}{48}$ in.

D. $\frac{1}{4}$ in.

16. Which is the correct number line representation of $\frac{3}{5} + \left(-\frac{2}{5}\right)$?

A.

B.

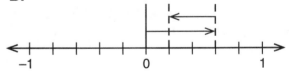

C.

D.

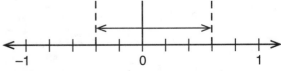

Name: _____ Date: _____

The Number System: Multiple-Choice Assessment Prep

Directions: Circle the choice that best answers the question.

---

17. Completely simplify the expression:

$$-\frac{5}{2} - (-9.3) + \left(-4\frac{1}{8}\right) + 6 - 7$$

   A. −16.925

   B. $1\frac{27}{40}$

   C. $-2\frac{3}{40}$

   D. 1.925

---

19. A clothing store holds a sale during which all items are 40% off their ticketed prices. Yuko purchases three shirts regularly priced $24.95 each and two sweaters regularly priced $42.50 each. How much must Yuko pay at the register after he uses a $25 gift card?

   A. $70.91

   B. $38.94

   C. $80.91

   D. $53.94

---

18. What is the simplified fractional form of $0.0\overline{6}$?

   A. $\frac{2}{3}$

   B. $\frac{2}{30}$

   C. $\frac{3}{50}$

   D. $\frac{1}{15}$

---

20. The quotient of $\frac{3}{10}$ and $\frac{2}{5}$ is decreased by the sum of $2\frac{1}{8}$ and $-7\frac{3}{4}$. That result is then multiplied by $\frac{8}{3}$. What is the final value?

   A. −13

   B. $18\frac{5}{9}$

   C. 17

   D. $-24\frac{1}{3}$

Name: _____ Date: _____

The Number System:
Open-Ended Response Assessment Prep

Directions: Answer the question completely. Show your work and explain your reasoning.

Problem 1: Your teacher asks you to illustrate the following problem on a number line:

$$0.6 + (-1) - (-0.8) - 0.4 = 0$$

a. Rewrite the problem using only addition.

b. Illustrate the problem on the given number line.

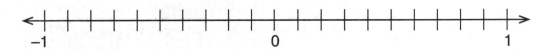

-1   0   1

c. Explain why the result of the calculations is zero.

Show your work.	Explain your reasoning.

Name: _____　Date: _____

The Number System:
Open-Ended Response Assessment Prep

Directions: Answer the question completely. Show your work and explain your reasoning.

Problem 2: A farmer leases $19\frac{1}{2}$ acres of his farmland to local residents wanting to grow their own organic vegetables. He divides this property into one-sixth-acre lots, each of which he rents for $345 per year, payable in equal monthly installments. How much monthly income does the farmer generate from leasing his farmland?

Show your work.	Explain your reasoning.

Name: _____ Date: _____

The Number System:
Open-Ended Response Assessment Prep

Directions: Answer the question completely. Show your work and explain your reasoning.

Problem 3: A large outdoor swimming pool is nearly full as the start of the summer season approaches. A fill pipe and garden hose are used simultaneously to fill the pool. Water flows through the fill pipe at a rate of 2,244 gallons per hour and the garden hose at a rate of $565\frac{4}{5}$ gallons per hour. The pool loses $112\frac{1}{2}$ gallons of water per hour due to evaporation. How long will it take to add the final $9\frac{1}{4}$ inches of water to the pool if it takes $7,654\frac{1}{2}$ gallons of water to raise the water level one inch?

Show your work.	Explain your reasoning.

Name: _____ Date: _____

The Number System:
Open-Ended Response Assessment Prep

Directions: Answer the question completely. Show your work and explain your reasoning.

Problem 4: The sum of $-7\frac{2}{5}$ and the product of 4.32 and $\frac{45}{16}$ is decreased by 64%. The quotient of that result and the difference between −11.74 and −7.18 is decreased by $1\frac{2}{3}$. Determine the simplified fractional form of the final value obtained.

Show your work.	Explain your reasoning.

The Number System: Answers and Diagnostics

Multiple Choice Questions

Problem #	Correct Answer	Diagnostics
1.	A	B. Student did not select fractional form of answer as directed. C. Student violated order of operations by adding before subtracting. D. Student did not reduce fraction.
2.	D	A. Student added rather than subtracting. B. Student subtracted whole numbers and fractions separately. C. Student incorrectly added denominators to whole numbers when converting mixed numbers to fractions.
3.	B	A. Student calculated correctly but labeled with incorrect compass direction. C. Student added rather than subtracting values. D. Student added values and labeled with incorrect compass direction.
4.	C	A. Student determined method to obtain $0.00 rather than starting amount. B. Student added values. D. Student misinterpreted lending or donating as a gain rather than a loss.
5.	B	A. Student omitted 0 in tenths place. C. Student concluded repeating block of digits was determined by first occurrence of a repeated digit rather than entire block of repeating digits. D. Student concluded decimal terminated when quotient of 0 was obtained.
6.	A	B. Student reported answer as fraction rather than decimal as directed. C. Student violated order of operations by multiplying before dividing. D. Student multiplied by $\frac{4}{9}$ rather than dividing.
7.	C	A. Student rounded up. B. Student multiplied rather than dividing. D. Student determined amount of brown sugar remaining after baking one pie.
8.	D	A. Student added. B. Student divided. C. Student subtracted.
9.	A	B. Student increased distance run each day by 15%. C. Student subtracted 0.15 rather than 15% from distance run each day. D. Student confused new routine and old routine.
10.	A	B. Student subtracted. C. Student averaged given values. D. Student did not understand the question.
11.	C	A. Student multiplied by denominator of complex fraction rather than dividing. B. Student multiplied by $\frac{3}{4}$ in denominator rather than dividing. D. Student entered expression on calculator using division to represent fraction bar and without using any parentheses.
12.	B	A. Student confused magnitudes of given electrical charges. C. Student interpreted given electrical charges as number of each type of ion needed to negate given charge for that ion. D. Student guessed.
13.	C	A. Student did not consider monthly payments for school. B. Student calculated correct value but responded with incorrect sign. D. Student did not consider monthly income.

The Number System: Answers and Diagnostics

14.	D	A. Student divided by scale factor rather than multiplying. B. Student added given values rather than subtracting. C. Student multiplied by 0.04 rather than 0.4 to calculate 40% of difference.
15.	D	A. Student divided 48 by 12 rather than 12 by 48. B. Student guessed. C. Student divided 1 inch into 48 parts rather than 12 inches into 48 parts.
16.	B	A. Student added $\frac{2}{5}$ rather than $-\frac{2}{5}$. C. Student began with $-\frac{3}{5}$ rather than $\frac{3}{5}$. D. Student graphed $\frac{3}{5}$ and $-\frac{2}{5}$ separately rather than in sequence.
17.	B	A. Student subtracted 9.3 rather than −9.3. C. Student performed all additions first, then all subtractions. D. Student added −4 and $\frac{1}{8}$ separately rather than as a single mixed number.
18.	D	A. Student ignored the 0 in the tenths place. B. Student determined correct value but did not reduce fraction as directed. C. Student did not recognize the 6 in the hundredths place repeated.
19.	A	B. Student calculated 40% of total price rather than 40% discount. C. Student deducted $25 gift card before calculating 40% discount. D. Student deducted $25 then calculated 40% of price rather than 40% discount.
20.	C	A. Student added sum of $2\frac{1}{8}$ and $-7\frac{3}{4}$ rather than subtracting. B. Student divided $\frac{2}{5}$ by $\frac{3}{10}$ rather than $\frac{3}{10}$ by $\frac{2}{5}$. D. Student subtracted $2\frac{1}{8}$ then added $-7\frac{3}{4}$ rather than subtracting their sum.

Open-Ended Response Questions

Problem #1

a. $0.6 + (-1) - (-0.8) - 0.4 = 0$ becomes $0.6 + (-1) + 0.8 + (-0.4) = 0$.

b.

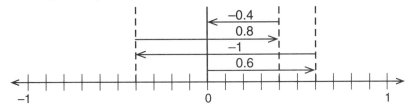

c. The sum of the positive values is $0.6 + 0.8 = 1.4$; the sum of the negatives is $-1 + (-0.4) = -1.4$.
The sum of opposite values, known as additive inverses (in this case, 1.4 and −1.4), is always zero.

Problem #2

$19\frac{1}{2} \div \frac{1}{6} = 117$ lots for lease; $117 \times 345 = \$40,365$ per year; $40,365 \div 12 = \$3,363.75$ per month
The farmer generates $3,363.75 in monthly income from leasing his farmland.

Problem #3

Amount of water added per hour: $2,244 + 565\frac{4}{5} - 112\frac{1}{2} = 2,697\frac{3}{10}$ gallons
Water needed: $7,654\frac{1}{2} \times 9\frac{1}{4} = 70,804\frac{1}{8}$ gallons; time to finish filling: $70,804\frac{1}{8} \div 2,697\frac{3}{10} = 26\frac{1}{4}$ hours
It will take $26\frac{1}{4}$ hours to finish filling the pool.

Problem #4

$4.32 \times \frac{45}{16} = 12.15$; $12.15 + (-7\frac{2}{5}) = 4.75$; $0.64 \times 4.75 = 3.04$; $4.75 - 3.04 = 1.71$
$-11.74 - (-7.18) = -4.56$; $1.71 \div (-4.56) = -0.375$; $-0.375 - 1\frac{2}{3} = -\frac{49}{24} = -2\frac{1}{24}$
The final value is $-\frac{49}{24}$ or $-2\frac{1}{24}$.

Expressions and Equations

Problem Correlation to CCSS Grade 7 Expressions and Equations Standards

MC Problem #	7.EE.A.1	7.EE.A.2	7.EE.B.3	7.EE.B.4	7.EE.B.4a	7.EE.B.4b
1					•	
2		•				
3	•					
4			•			
5						•
6			•			
7						•
8	•					
9					•	
10					•	
11			•			
12		•				
13			•			
14						•
15			•			
16					•	
17				•		
18						•
19			•			
20					•	
Open-Ended Problem #	7.EE.A.1	7.EE.A.2	7.EE.B.3	7.EE.B.4	7.EE.B.4a	7.EE.B.4b
1					•	
2	•					
3			•			
4						•

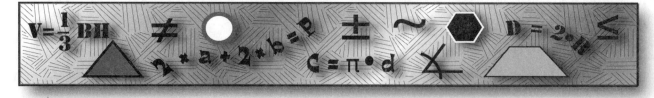

Name: _____   Date: _____

Expressions and Equations: Multiple-Choice Assessment Prep

Directions: Circle the choice that best answers the question.

---

1. Determine the width of a rectangle with a perimeter of 84 inches and a length of 23 inches.

   A. 19 in.

   B. 61 in.

   C. 38 in.

   D. 30.5 in.

---

3. What is the correct simplified form of the given expression?

$$\frac{1}{2}(5x - 6) + \frac{1}{3}$$

   A. $\frac{5}{2}x - \frac{10}{3}$

   B. $\frac{5}{2}x - \frac{17}{3}$

   C. $\frac{5}{2}x - \frac{17}{6}$

   D. $\frac{5}{2}x - \frac{8}{3}$

---

2. Let p represent the price of a toy at a toy store. Sales tax is 6%. Which of the following is a correct expression for the total amount paid at the register?

   A. $p + 0.06$

   B. $1.6p$

   C. $1.06p$

   D. $0.06p$

---

4. Mei checked a book out of the public library two days ago, and she read half the book that night. Yesterday, she read 40% of the book. What fraction of the book does Mei have left to read today?

   A. $\frac{1}{4}$

   B. $\frac{3}{10}$

   C. $\frac{1}{10}$

   D. $\frac{9}{10}$

Name: _____   Date: _____

Expressions and Equations: Multiple-Choice Assessment Prep

Directions: Circle the choice that best answers the question.

---

5. Jose has a total of $96 to spend on baseball cards for his collection and a catcher's mitt for his Little League season. Each pack of baseball cards costs $2.50, and the catcher's mitt he wants costs $60. What is the greatest number of packs of baseball cards Jose can buy so that he can still afford the catcher's mitt he wants?

A. 14

B. 15

C. 62

D. 72

---

7. Cindy wants to buy a new road bike that costs $2,199. She currently has $800 saved for the bike, and she can add $150 each month to her bike fund. How many months will it take for Cindy to save enough to buy the new bike?

A. 9

B. 20

C. 15

D. 10

---

6. A cell phone that normally costs $270 is on sale for one-third off the original price. Sales tax is 5%. What is the final amount paid at checkout?

A. $185.00

B. $189.00

C. $94.50

D. $270.00

---

8. Determine which of the expressions is equivalent to the following:

$$4(x + 3) - x + 9$$

A. $3(x + 4)$

B. $3x + 3$

C. 25

D. $3(x + 7)$

Name: _____　Date: _____

Expressions and Equations: Multiple-Choice Assessment Prep

Directions: Circle the choice that best answers the question.

9. A gas cylinder contains $3\frac{1}{4}$ cubic feet of helium. Each balloon to be inflated holds $\frac{1}{16}$ cubic foot of helium. How many balloons would need to be inflated to reduce the volume of helium left in the cylinder to $1\frac{7}{8}$ cubic feet?

   A. 82

   B. 36

   C. 22

   D. 52

10. A pot of water had an initial temperature of 52°C. The boiling point of water is 100°C. The pot has been on the stove long enough to heat the water to a temperature that is 65% of the way from its original temperature to its boiling point. What is the current temperature of the water?

   A. 31.2°C

   B. 83.2°C

   C. 58.8°C

   D. 65.0°C

11. Taro has a savings fund he uses to go on tropical vacations. He first books a trip to Aruba that uses two-fifths of the savings fund. Then he books a trip to Jamaica that uses three-tenths of the remaining amount. What percent of the beginning balance of the savings fund is left after booking both trips?

   A. 42%

   B. 30%

   C. 12%

   D. 78%

12. Let n represent the original number of students competing in a spelling bee. After the first round of competition, the number of students is $0.9n$. Choose the best description of what happened during the first round of competition from the choices below.

   A. Ninety percent more students joined the competition.

   B. Ten percent of the original competitors were eliminated.

   C. Nine-tenths of the original competitors were eliminated.

   D. Ninety-one percent of the original competitors remain in the competition.

Name: _____ Date: _____

Expressions and Equations: Multiple-Choice Assessment Prep

Directions: Circle the choice that best answers the question.

---

13. Greta has a bag containing 64 marbles. She first gives 8 marbles each to Robin and Ramona. Then, Greta gives half of the remaining marbles to Ling. Finally, 25% of the marbles Greta has left go to Maeko. How many marbles does Greta still have?

A. 6

B. 21

C. 12

D. 18

---

15. Yolanda throws the discus 72 feet at her first track meet of the season. Her throw at the second meet is 25% longer. She throws the discus $\frac{9}{10}$ as far at the third track meet as she does at the second. How much farther does Yolanda throw the discus at her third track meet than at her first?

A. 28 ft

B. 9 ft

C. 81 ft

D. Yolanda did not throw the discus as far at her third track meet as at her first.

---

14. A plant is currently $2\frac{3}{8}$ inches tall and is growing at a rate of $\frac{3}{4}$ inch per day. After how many days will the plant reach a height that exceeds 10 inches?

A. 10

B. 6

C. 11

D. 17

---

16. A building constructed on swamp land is sinking at a rate of 0.018 meter each year. The top of the building is initially 12.986 meters above ground. After how many years will the top of the building be only 12.5 meters above ground?

A. 0.468

B. 27

C. 2.7

D. 714.5

Name: _____   Date: _____

Expressions and Equations: Multiple-Choice Assessment Prep

Directions: Circle the choice that best answers the question.

17. Determine which equation below correctly represents the following: ten less than three times the sum of x and seven is twenty-six.

    A. $3(x + 7) - 10 = 26$

    B. $3x + 7 - 10 = 26$

    C. 5

    D. $10 - 3(x + 7) = 26$

18. A school schedules classes so students receive 138 hours of mathematics instruction each year. Every day a student misses math class, that student loses three-fourths of an hour of mathematics instruction. State law requires students to receive a minimum of 124 hours of mathematics instruction to be able to complete a math course. How many days can a student miss math class and still be able to complete the course?

    A. 18

    B. 10

    C. 19

    D. 11

19. Chen broke his leg in a skiing accident. While recovering, he gained 8% of his weight because of not being able to exercise. Once healed, he began exercising and lost 22 pounds. Chen weighed 162.5 pounds before the accident. How much does he weigh now?

    A. 148.5 lbs

    B. 127.5 lbs

    C. 136.89 lbs

    D. 153.5 lbs

20. A number is increased by $\frac{2}{5}$. Multiplying that sum by 20 gives 98. What is the number?

    A. $\frac{53}{10}$

    B. $\frac{388}{5}$

    C. $\frac{9}{2}$

    D. $\frac{122}{25}$

Name: _____ Date: _____

Expressions and Equations: Open-Ended Response Assessment Prep

Directions: Answer the question completely. Show your work and explain your reasoning.

Problem 1: Elena and Isabel are standing back to back at center court in a basketball arena. They take 20 paces away from one another. When they stop, they are exactly 30 meters apart. Each of Elena's paces measures 0.77 meter. How long is each of Isabel's paces?

Show your work.	**Explain your reasoning.**

Name: _____ Date: _____

Expressions and Equations: Open-Ended Response Assessment Prep

Directions: Answer the question completely. Show your work and explain your reasoning.

Problem 2: An outdoor swimming pool has a width of *w* feet. Its length is 130% of its width. The pool is surrounded by a four-foot wide concrete deck. Write and then completely simplify an expression for the perimeter of the outer edge of the concrete deck.

Show your work.	Explain your reasoning.

Name: _____ Date: _____

Expressions and Equations: Open-Ended Response Assessment Prep

Directions: Answer the question completely. Show your work and explain your reasoning.

Problem 3: Juanita wants to hang a picture in the center of a wall that measures 187 inches across. The picture measures $31\frac{3}{4}$ inches across. How far from the left edge of the wall should the left edge of the picture be?

Show your work.	Explain your reasoning.

Name: _____ Date: _____

Expressions and Equations: Open-Ended Response Assessment Prep

Directions: Answer the question completely. Show your work and explain your reasoning.

Problem 4: Marcus works at a sales job that pays $1,750 per month plus $150 per sale. The fishing boat he wants to buy costs $4,200. How many sales must Marcus make this month to earn enough to purchase the boat? Write an inequality to solve the problem, and describe the solutions.

Show your work.	Explain your reasoning.

Expressions and Equations: Answers and Diagnostics

Multiple-Choice Questions

Problem #	Correct Answer	Diagnostics
1.	A	B. Student only subtracted one length. C. Student did not divide by two to obtain one width. D. Student did not subtract two lengths.
2.	C	A. Student did not multiply sales tax by original price. B. Student incorrectly converted 6% to 0.6. D. Student did not add original price to sales tax to obtain final amount.
3.	D	A. Student subtracted $\frac{1}{3}$. B. Student did not distribute $\frac{1}{2}$ to −6. C. Student multiplied $\frac{1}{3}$ by $\frac{1}{2}$.
4.	C	A. Student incorrectly converted 0.4 (decimal form of 40%) to $\frac{1}{4}$. B. Student calculated 40% of remaining half rather than 40% of book. D. Student responded with fraction read rather than fraction remaining.
5.	A	B. Student rounded up. C. Student added 60 and 96. D. Student used incorrect order of operations.
6.	B	A. Student added $5 tax rather than 5%. C. Student used $\frac{1}{3}$ of original price as sale price rather than discount. D. Student used 0.5 rather than 0.05 for sales tax.
7.	D	A. Student rounded down rather than going up to next integer value, as indicated by the context. B. Student added $800 to $2,199. C. Student did not account for the $800 already saved.
8.	D	A. Student did not distribute properly. B. Student subtracted nine instead of adding. C. Student obtained four when subtracting x from $4x$.
9.	C	A. Student added $3\frac{1}{4}$ and $1\frac{7}{8}$. B. Student guessed. D. Student calculated total number of balloons that can be inflated.
10.	B	A. Student did not add amount of increase to original temperature. C. Student subtracted amount of increase from boiling point. D. Student calculated 65% of boiling point.
11.	A	B. Student subtracted $\frac{3}{10}$ of original amount for second trip. C. Student multiplied $\frac{2}{5}$ by $\frac{3}{10}$ and did not subtract from the original amount. D. Student added $\frac{3}{10}$ of amount after first trip rather than subtracting.
12.	B	A. Student interpreted $0.9n$ as an increase, not final amount. C. Student interpreted $0.9n$ as amount of decrease, not final amount. D. Student interpreted 0.9 as 9%.
13.	D	A. Student calculated the final 25% to be given away, but did not subtract it to see how many marbles would be left. B. Student only subtracted 8 once. C. Student used half the original amount, not half after subtracting 8 each to be given to Robin and Ramona.

Expressions and Equations: Answers and Diagnostics

14.	C	A. Student rounded down rather than going up to the next integer value, as indicated by the context. B. Student multiplied by $\frac{3}{4}$ instead of $\frac{4}{3}$. D. Student added $2\frac{3}{8}$.
15.	B	A. Student multiplied by $\frac{10}{9}$. C. Student responded with final distance, not difference from original. D. Student used 25% of original value rather than adding it to original.
16.	B	A. Student added 0.018 rather than dividing by it. C. Student divided by -0.18 rather than -0.018. D. Student multiplied left side of equation ($12.986 - 0.018x = 12.5$) by 1,000 but right side by only 10 to obtain integer coefficients.
17.	A	B. Student did not properly express the sum of x and 7 as a quantity. C. Student solved for x rather than just writing the equation. D. Student subtracted from 10 rather than subtracting 10.
18.	A	B. Student multiplied by $\frac{3}{4}$. C. Student rounded up rather than down or did not reverse inequality. D. Student multiplied by $\frac{3}{4}$ and rounded up instead of down.
19.	D	A. Student added eight pounds rather than 8% of original weight. B. Student subtracted 8% of original weight. C. Student subtracted 22% rather than 22 pounds.
20.	C	A. Student added eight. B. Student subtracted 20 rather than dividing by 20 without distributing. D. Student did not distribute 20 properly.

Open-Ended Response Questions

Problem #1
20 combined paces (i.e., 20 times total distance covered by one of each girl's paces) gives total of 30 m
$20(x + 0.77) = 30 \rightarrow 20x + 15.4 = 30 \rightarrow 20x = 14.6 \rightarrow x = 0.73$
The length of Isabel's pace is 0.73 m.

Problem #2
Length of pool: $1.3w$ Outer width of concrete deck: $w + 8$ Outer length of concrete deck: $1.3w + 8$
Outer perimeter of concrete deck: $2(w + 8) + 2(1.3w + 8) = 2w + 16 + 2.6w + 16 = 4.6w + 32$
The outer perimeter of the concrete deck is $4.6w + 32$ feet.

Problem #3
Center of wall: $187 \div 2 = 93\frac{1}{2}$ or 93.5 inches from left edge of wall

Center of picture: $31\frac{3}{4} \div 2 = 15\frac{7}{8}$ or 15.875 inches from left edge of picture

$93\frac{1}{2} - 15\frac{7}{8} = 77\frac{5}{8}$ inches, or $93.5 - 15.875 = 77.625$ in.

The left edge of the picture should be $77\frac{5}{8}$ or 77.625 inches from the left edge of the wall.

Problem #4
Let x represent number of sales; $1,750 + 150x \geq 4,200 \rightarrow 150x \geq 2,450 \rightarrow x \geq 16.\overline{3}$
Number of sales must be at least $16.\overline{3}$, but must be a whole number, so round up to 17.
Marcus must make at least 17 sales this month to earn enough to purchase the boat.

Statistics and Probability

Problem Correlation to CCSS Grade 7 Statistics and Probability Standards

MC Problem #	7.SP.A.1	7.SP.A.2	7.SP.B.3	7.SP.B.4	7.SP.C.5	7.SP.C.6	7.SP.C.7	7.SP.C.7a	7.SP.C.7b	7.SP.C.8	7.SP.C.8a	7.SP.C.8b	7.SP.C.8c
1					•								
2	•												
3						•							
4		•											
5				•									
6								•					
7													•
8												•	
9											•		
10					•								
11									•				
12												•	
13					•								
14									•				
15						•							
16									•				
17											•		
18		•											
19	•												
20				•									
Open-Ended Problem #	7.SP.A.1	7.SP.A.2	7.SP.B.3	7.SP.B.4	7.SP.C.5	7.SP.C.6	7.SP.C.7	7.SP.C.7a	7.SP.C.7b	7.SP.C.8	7.SP.C.8a	7.SP.C.8b	7.SP.C.8c
1		•											
2				•									
3								•			•	•	

Name: _____   Date: _____

Statistics and Probability: Multiple-Choice Assessment Prep

Directions: Circle the choice that best answers the question.

1. A random event is likely to occur. Which value could be the probability associated with the event?

   A. 0.83

   B. 110%

   C. $\frac{1}{2}$

   D. −0.5

2. A survey is conducted to determine the favorite sports of all students attending ABC High School. Which process of selecting a sample of students will have the best chance of producing valid results?

   A. Ask all the girls in the school.

   B. Survey the entire senior class.

   C. Select all students taking advanced math.

   D. Write the name of each student in the school on a separate piece of paper, shuffle all the papers, randomly select 100 papers, and ask those students.

3. Ana rolls a number cube 6,000 times. Which is the best description of the number of times Ana should expect to roll a five?

   A. Exactly 1,000 times

   B. Approximately 1,000 times

   C. More than 1,000 times

   D. Less than 1,000 times

4. The lifespans, in days, of a random sample of 30 houseflies are provided. Use the data to determine the average lifespan of a housefly.

   17 18 19 20 21 24 24 24 24 24
   24 24 24 25 26 26 26 26 26 27
   27 27 27 28 28 28 28 29 29 30

   A. 24 days

   B. 25 days

   C. 26 days

   D. None of the above

Statistics and Probability: Multiple-Choice Assessment Prep

Directions: Circle the choice that best answers the question.

5. The mean age of evergreens in Ozark Pine Forest is 18 years less than the mean age of evergreens in Appalachia Pine Forest. The mean absolute deviation of ages is eight years for both forests. Which must be a true statement about the two distributions of ages when graphed on a dot plot?

A. The graph for Ozark Pine Forest is less spread out.

B. The graph for Appalachia Pine Forest has taller columns.

C. The separation between the two distributions of ages is noticeable.

D. All of these statements are true.

6. Xavier is one member of a soccer team consisting of 23 players. After a game, one member of the team is randomly selected for a radio interview. What is the probability Xavier is not selected for the interview?

A. $\frac{22}{23}$

B. $\frac{1}{23}$

C. 23%

D. None of the above

7. Single random digits, 0 through 9, are used to simulate selection of individuals from a population in which 40% of people are fluent in two languages. Which interpretation of the random digits is valid for the simulation?

A. A prime number represents a person fluent in two languages, but no other digit does.

B. Obtaining a 0, 1, 2, or 3 indicates an individual fluent in two languages is selected; obtaining any other digit does not.

C. A single-digit factor of 45 corresponds to an individual fluent in two languages; any other digit corresponds to someone who is not.

D. All of the above are valid.

8. Which is the sample space for flipping a coin three times in order to determine the probability of flipping three tails?

A. TTT

B. HHT, HTT, THT, TTT

C. HHH, HHT, HTH, HTT, THH, THT, TTH, TTT

D. T, TT, TTT

Name: _____   Date: _____

Statistics and Probability: Multiple-Choice Assessment Prep

Directions: Circle the choice that best answers the question.

---

9. Naomi randomly guesses the answers to a True/False question followed by a multiple-choice question with four possible answers. What is the probability Naomi answers both questions correctly?

A. $\frac{3}{4}$

B. $\frac{1}{4}$

C. $\frac{1}{8}$

D. $\frac{1}{2}$

---

11. Determine the probability of spinning an odd number on the spinner shown.

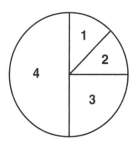

A. 25%

B. 37.5%

C. 50%

D. 62.5%

---

10. The heights, in centimeters, of the players on two rugby sevens teams are shown. Determine which team has less variability in height based on mean absolute deviation.

Team I: 170, 178, 177, 163, 158, 183, 182

Team II: 183, 190, 178, 189, 185, 190, 180

A. Team I

B. Team II

C. Neither

D. Cannot be determined

---

12. Leon randomly selects two blocks from a bag containing three red and five yellow blocks. Use the sample space provided to identify which outcomes describe the event in which Leon selects one red and one yellow block.

S = { RR, RY, YR, YY }

A. RR, YY

B. RY

C. RR, RY, YR, YY

D. RY, YR

Statistics and Probability: Multiple-Choice Assessment Prep

Directions: Circle the choice that best answers the question.

---

13. The probability of a particular event occurring is $\frac{3}{1000}$. Which statement best characterizes the event?

 A. The event is not likely to occur.

 B. The event is likely to occur.

 C. The event is certain to occur.

 D. It is impossible for the event to occur.

---

15. In a study of 300 randomly selected university students, 186 are in-state residents. Based on this study, how many of the 9,850 students attending the university would one expect to be out-of-state residents?

 A. 3,743

 B. 6,107

 C. 114

 D. 38

---

14. Determine the probability that a randomly selected month has less than 31 days.

 A. $\frac{1}{3}$

 B. $\frac{1}{12}$

 C. 1

 D. $\frac{5}{12}$

---

16. One day of the week is selected at random. To the nearest tenth of a percent, what is the probability that the chosen day is Saturday?

 A. 50.0%

 B. 28.6%

 C. 14.3%

 D. 85.7%

Name: _____   Date: _____

Statistics and Probability: Multiple-Choice Assessment Prep

Directions: Circle the choice that best answers the question.

---

17. A single digit is chosen at random. What is the probability the digit is prime or a multiple of three?

A. $\frac{7}{10}$

B. $\frac{3}{5}$

C. $\frac{2}{3}$

D. $\frac{7}{9}$

19. In a survey of 100 randomly selected pet owners, 59 own at least one dog. Based on the results of the survey, which generalization about all pet owners is most valid?

A. Forty-one percent of all pet owners own more than one dog.

B. One hundred people own dogs.

C. Fifty-nine percent of all pet owners own at least one dog.

D. No valid conclusions can be drawn about all pet owners since only 100 were surveyed.

---

18. Shen estimates there are 500 kernels per ear of corn. He calculates the mean and mean absolute deviation (MAD) for three different samples of 100 ears each. Based on the given results, which statement is correct?

	Mean	MAD
Sample 1	491.38	15.72
Sample 2	504.66	10.08
Sample 3	512.72	19.46

A. Shen's estimate appears to be valid.

B. Shen's estimate appears too high.

C. Shen's estimate appears too low.

D. No valid conclusion can be drawn.

20. Eight 5th-grade and eight 7th-grade mathematics textbooks are randomly selected. Their weights, in pounds, are shown. Based on this data, which statement regarding the average weights of 5th-grade and 7th-grade math texts is most valid?

Grade 5: 2.2, 2.4, 2.5, 3.7, 3.7, 3.8, 3.8, 3.9
Grade 7: 3.1, 3.2, 3.2, 3.2, 3.2, 3.3, 3.4, 3.4

A. Math texts for 7th grade weigh more.

B. Math texts for 5th grade weigh more.

C. Math texts for 5th and 7th grades are heavy.

D. There is no noticeable difference in average weights of math texts in 5th and 7th grades.

Name: _____ Date: _____

Statistics and Probability: Open-Ended Response Assessment Prep

Directions: Answer the question completely. Show your work and explain your reasoning.

Problem 1: A random sample of 36 honeybees is collected. The weights, in milligrams, of the bees are given. Use this data to predict to the nearest percent how many honeybees weigh more than one mean absolute deviation above the mean honeybee weight.

102	103	94	105	88	101	95	98	99	92	103	99
103	95	96	100	90	104	91	92	99	102	95	103
101	101	93	96	104	102	98	92	92	100	97	103

Show your work.

Explain your reasoning.

Name: _____　Date: _____

Statistics and Probability: Open-Ended Response Assessment Prep

Directions: Answer the question completely. Show your work and explain your reasoning.

Problem 2: The number of rainy days per year in London (L) and New York (NY) are recorded over a 15-year period. Use the given data to answer the questions that follow.

L	111	121	118	104	100	112	106	110
	116	102	113	104	107	115	114	
NY	107	135	119	125	109	136	120	122
	132	116	102	134	129	119	122	

a. On average, which city experiences more rainy days per year, and by how much?

b. Based on mean absolute deviation, which city experiences more variation in rainy days per year?

Show your work.	Explain your reasoning.

Name: _____ Date: _____

Statistics and Probability: Open-Ended Response Assessment Prep

Directions: Answer the question completely. Show your work and explain your reasoning.

Problem 3: Consider an experiment consisting of two parts: 1) randomly select one of the vowels in the English alphabet; and 2) roll a number cube.

a. Give the sample space for the experiment, and determine the number of outcomes in the sample space.

b. Determine the probability that the letter A is selected followed by rolling a one.

c. Determine the probability that the letter O is selected.

d. Determine the probability that a six is rolled on the number cube.

e. Determine the probability that the letter U is selected or a four is rolled.

f. Determine the probability that a vowel in the word PRIME is selected and a prime number is rolled.

Show your work.	Explain your reasoning.

Statistics and Probability: Answers and Diagnostics

Multiple-Choice Questions

Problem #	Correct Answer	Diagnostics
1.	A	B. Student incorrectly selected value greater than 100%. C. Student did not understand concept of "likely to occur." D. Student did not understand probabilities must range from 0 to 1.
2.	D	A. Student did not consider that male and female opinions may differ greatly. B. Student did not consider that opinions may differ based on student age. C. Student did not consider that students in similar academic classes may not represent entire student body.
3.	B	A. Student determined exact theoretical value but did not account for error. C. Student provided optimistic result but did not account for error. D. Student provided pessimistic result but did not account for error.
4.	B	A. Student determined mode rather than mean. C. Student determined median rather than mean. D. Student guessed.
5.	C	A. Student did not understand concept of mean absolute deviations being equal. B. Student assumed greater mean corresponds to taller columns on dot plot. D. Student incorrectly identified incorrect statements as correct.
6.	A	B. Student determined probability of Xavier being selected. C. Student guessed. D. Student did not understand the problem.
7.	D	A. Forty percent of the 10 single digits (0 through 9) are prime: 2, 3, 5, and 7. B. The digits 0, 1, 2, and 3 make up 40% of the 10 single digits (4 of the 10). C. The 4 single-digit factors of 45 (1, 3, 5, and 9) make up 40% of the single digits.
8.	C	A. Student determined only the successful outcome, not entire sample space. B. Student determined all outcomes in which tail occurs on third coin flip. D. Student did not understand concept of sample space.
9.	C	A. Student added probabilities for each question rather than multiplying. B. Student used $\frac{1}{2}$ as probability for answering multiple-choice question correctly. D. Student considered both questions collectively as simply correct or incorrect.
10.	B	A. Student determined smaller mean rather than mean absolute deviation. C. Student guessed. D. Student did not understand the question.
11.	B	A. Student guessed. C. Student incorrectly assumed all outcomes have equal probability of occurring. D. Student determined probability of spinning an even number.
12.	D	A. Student did not understand the question. B. Student did not consider selection of yellow before red. C. Student listed entire sample space rather than only desired outcomes.
13.	A	B. Student confused concepts of likely and unlikely events. C. Student did not understand concept of certain event. D. Student assumed extremely small probability implies an impossible event.
14.	D	A. Student did not consider all 5 desired months (Feb., Apr., Jun., Sep., Nov.). B. Student calculated probability of selected month having less than 30 days. C. Student calculated probability of month having less than or equal to 31 days.
15.	A	B. Student calculated the expected number of in-state residents. C. Student calculated desired value for sample rather than population. D. Student calculated percent of out-of-state residents.

Statistics and Probability: Answers and Diagnostics

16.	C	A. Student considered weekend days (Sat. and Sun.) to be entire sample space. B. Student calculated probability of selecting either weekend day. D. Student calculated probability of not selecting Saturday.
17.	B	A. Student counted the digit 3 twice (as both prime and a multiple of 3). C. Student did not consider 0 as a digit, giving a sample space of only 9 digits. D. Student counted the digit 3 twice and did not consider 0 as a digit.
18.	A	B. Student did not interpret data correctly. C. Student added mean and mean absolute deviation for each sample. D. Student did not interpret data correctly.
19.	C	A. Student misinterpreted survey results. B. Student did not understand nature of random sample collected. D. Student did not understand concept of generalizing through random samples.
20.	D	A. Student calculated at least one of the means incorrectly. B. Student calculated at least one of the means incorrectly. C. Student provided opinion rather than mathematically valid conclusion.

Open-Ended Response Questions

Problem #1

Mean: 98 mg; mean absolute deviation (MAD): 4 mg; one MAD above mean: $98 + 4 = 102$ mg
Number of data values above 102 mg: 8; total number of data values: 36; $8 \div 36 = 0.\overline{2} \approx 22\%$
22% of honeybees weigh more than one mean absolute deviation above the mean honeybee weight.

Problem #2

  a. London average: 110.2 days; New York average: 121.8 days; difference: $121.8 - 110.2 = 11.6$ days
     New York experiences 11.6 more rainy days per year.
  b. London MAD: 5.12 days; New York MAD: 8.08 days
     New York experiences more variation in rainy days per year.

Problem #3

  a. Multiple formats exist; sample format as follows:
     { A1, A2, A3, A4, A5, A6, E1, E2, E3, E4, E5, E6, I1, I2, I3, I4, I5, I6, O1, O2, O3, O4, O5, O6, U1, U2, U3, U4, U5, U6 }; The number of outcomes in the sample space is 30.

  b. Outcome matching criteria: A1 (1 outcome); Desired probability: $\frac{1}{30}$

     The probability of selecting the letter A followed by rolling a one is $\frac{1}{30}$.

  c. Outcomes matching criteria: O1, O2, O3, O4, O5, O6 (6 outcomes)
     Desired probability: $\frac{6}{30} = \frac{1}{5}$; The probability of selecting the letter O is $\frac{1}{5}$.

  d. Outcomes matching criteria: A6, E6, I6, O6, U6 (5 outcomes)
     Desired probability: $\frac{5}{30} = \frac{1}{6}$; The probability of rolling a six is $\frac{1}{6}$.

  e. Outcomes matching criteria: A4, E4, I4, O4, U1, U2, U3, U4, U5, U6 (10 outcomes)
     Desired probability: $\frac{10}{30} = \frac{1}{3}$; The probability of selecting a U or rolling a four is $\frac{1}{3}$.

  f. Outcomes matching criteria: E2, E3, E5, I2, I3, I5 (6 outcomes)
     Desired probability: $\frac{6}{30} = \frac{1}{5}$; The probability of selecting a vowel in the word PRIME and rolling a prime number is $\frac{1}{5}$.